Design Ideas For Sm[illegible]

Bedrooms & Baths

Copyright © 1996 by Rockport Publishers, Inc.

All rights reserved. No part of this book may be reproduced in any form without written permission of the copyright owners. All images in this book have been reproduced with the knowledge and prior consent of the artists concerned and no responsibility is accepted by producer, publisher, or printer for any infringement of copyright or otherwise, arising from the contents of this publication. Every effort has been made to ensure that credits accurately comply with information supplied.

First published in the United States of America by:
Rockport Publishers, Inc.
146 Granite Street
Rockport, Massachusetts 01966-1299
Telephone: (508) 546-9590
Fax: (508) 546-7141

ISBN 1-56496-302-0

10 9 8 7 6 5 4 3 2 1

Layout: Sara Day Graphic Design
Cover Credit: See page 27

Printed by Welpac, Singapore.

DESIGN IDEAS FOR SMALL SPACES

Bedrooms & Baths

Norman Smith

Rockport Publishers
Rockport, Massachusetts

INTRODUCTION

Small space design is about making a lot out of a little. No matter what the location, type of residence or price range, our residential environments have to work harder than ever before, becoming more efficient and more accommodating in a smaller physical envelope.

By definition, small spaces have problems that are a direct result of their size. Too little wall space, too few windows and not enough natural light are drawbacks in any room—but they can be particularly troublesome in a small space. Likewise, poor ventilation, crowded furniture layout, and lack of adequate clothes storage are common problems in small bedrooms and baths. A poor design is magnified in a small space: A room that looks cramped or cluttered, dark, unappealing, and at times even oppressive is without a doubt, an ineffective and unsuccessful space.

Small spaces can be found anywhere; in urban apartments or condominiums, as well as in suburban, single-family houses. A small space can be a bedroom not much larger than the bed, or a room that combines space for study with sleeping arrangements for an infant and parents. It might be a niche-like hideaway carved from another room, or a tiny studio bedroom closet that uses built-in storage to save floor space. It might also be a tiny loft space that provides just enough floor area for a single bed tucked out of the way, or a piece of custom furniture that accommodates living and sleeping while acting as a room divider.

One way to make a small bedroom or bath appear larger and more gracious is to incorporate the room with an outdoor area—such as a deck or balcony. Using trim or strategically placed lighting and views to fool the eye and distract attention from the limited amount of space can also work well. In the same way, adding a steeply pitched cathedral ceiling (or creating the illusion of one) can make a less than grand bedroom appear larger and more gracious.

The purpose of this book is to explore small space design by illustrating the many individual concepts and approaches of designers and architects confronted with the problems of small spaces. The bedrooms and baths in these pages employ a variety of design tools or 'devices'. As you will see, while the devices appear again and again, their final effect changes according to the designer's choices, the use of the space, and the many other factors that go into a finished design.

From a designer's standpoint, a small space is never and should never be, a lost cause. By the same token, whether you plan to hire a designer or do the work yourself, no small space is beyond hope or impossible to improve. Whatever the problem you want to solve, any room or space in your home can be designed to appear more expansive and more attractive.

DESIGN, AESTHETICS, AND STYLE

Photographer: Walter Smalling

A well-built design project must be detailed and constructed properly.

Creating a good design plan is a lot like sculpting a figure or painting a picture. Then, that thought is translated from the mind to reality in the same way that a sculptor begins with a block of stone or a painter moves to the canvas. Finally, the initial concept is refined using the tools of each person's trade. For a sculptor, a hammer and chisels bring the idea to light, while for the painter, brushes and pigments suffice. For the designer, the essential concept is first tempered by the need to satisfy three basic qualities. Paraphrasing an ancient architectural theorist, all designs must be well-built and useable while incorporating a strong measure of delight.

Each of these basics is related to all the others so that when artfully conceived, the whole is a wonderful sum of these parts. The loss of any one of these qualities or an imbalance among them can make a design lackluster and unusable.

The design starts with an essential concept much like a thought or a wish. In small space designs, this care is manifested by the artful detailing of portions of the design and by the efficient, imaginative, and conscientious use of materials. Sensitivity to the interior and exterior environment is an important part of building. Simultaneously, the manipulation of a room's proportions must come into play for both aesthetic effect and ease in construction. The use of symmetry or, alternately, asymmetry in the structural layout affects the cost of a design and its feeling of openness.

Along with the sense of construction, the design must be useable and must satisfy basic functional requirements. In a small space, this can be as simple as increasing light and air and as complicated as integrating multiple uses in a tightly defined room. Again, a designer's sense of appropriate proportions must come into play to help shape the space effectively.

Just as proportions affect the feel of a room, the use of one or more focal points can create the appearance of spaciousness or alternately that of a cozy and enclosed retreat within the space. Spatial and depth perception can be manipulated to visually transform a small space. For instance, a wall can be angled to counteract the natural shortening of perspective, making a room seem larger or longer, while still accommodating required traffic patterns through the space. The use of symmetry to balance competing elements can lend an air of quiet repose to a small room while providing for an efficient use of the space. In a different situation, an asymmetrical layout might help counter unequal uses and impart a feeling of dynamic excitement to an otherwise plain room. Continuity and discontinuity are also often associated with small space design. Making a surface discontinuous reinforces a sense of separation; on the other hand, a continuous patch of unbroken floor can be used to tie together several adjacent small spaces so that each appears a part of the others.

Delight is the third part of the essential triumvirate but it is of a piece with all the rest. Delight is often confused with aesthetics and style. However, aesthetics is really just the pursuit and appreciation of beauty in its many forms. Various styles are, and have become over time, the accepted manifestation of a current standard of beauty. However, the phrase 'aesthetics and style' is often used as though they have a life of their own and are merely a window dressing that can be applied to any room, almost like throwing on a change of clothes. Nothing could be further from the truth.

From a designer's standpoint the overall atmosphere of a space, or its aesthetic, is not simply a question of overlaying a style. Rather, it is dictated by the interaction of all three essential elements. These requirements are then filtered through the designer's mind and mixed with the variables of budget and other requirements to produce a final aesthetic that is appropriate for each particular situation.

Finally, the design is further refined and made concrete with the tools or devices of the designers trade to produce a final product. These devices are part and parcel of every small space design and, in different ways, will always reflect the three essential qualities and their related concepts.

Photographer: Walter Smalling

While it is possible to redo rooms in different period styles to pleasurable effect, the best small space designs don't start with a style but rather with the designer's and owner's fundamental thoughts and needs.

BASIC CONCEPTS OF SMALL SPACE DESIGN

A design is shaped by the designer's knowledge of how these qualities are first realized. This occurs through the use of the designer's equivalent of canvas or stone, using several basic design concepts including:

SYMMETRY/ASYMMETRY refers to a plan configuration or to the three-dimensional treatment of surfaces and planes within a space. In an effort to resolve fundamental building and design problems, designers have begun experimenting with more extreme forms of asymmetry. It is frequently used in small space design for exactly this reason.

PROPORTION is essentially the relationship of parts to the whole both in two dimensions and in three. Certain proportional relationships are pleasing to the mind and eye while others can create a disharmonious appearance.

FOCUS/FOCAL POINTS can be one of the basic building blocks of any small space design. Creating a main focal point in a small space can minimize other unattractive but necessary intrusions in the space. A focal point can be many things; a painting on a wall, the wall itself, a superb cabinet, a view outside, or any number of other items.

SPATIAL/DEPTH PERCEPTION relates to the overall appearance of a space. A skillful designer can deliberately manipulate elements of the design to make the room feel larger or smaller, shorter or taller, narrower or wider, depending on the situation.

CONTINUITY/DISCONTINUITY applies to the nature of a surface or to the layout of a plan. Continuity usually melds or ties together disparate elements; discontinuity heightens differences to create an appealing tension.

As a design progresses, these qualities and concepts are combined in any number of ways to refine the initial thought.

HOW TO USE THIS BOOK

For the purposes of this book, eight different devices have been chosen to help explain the different approaches of the various small space designs and their respective designers.

While no design will employ every possible device and some will certainly employ more than just a few, it's still possible to look at small space design as a compendium of these tools and to imagine how these devices might be applied to other designs, including your own.

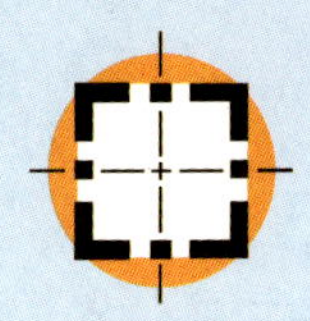

Plan Organization

Basic plan devices include:

a. A layout of one or more spaces that incorporates circulation or alternately, that reconfigures a space to remove circulation in order to make the space more useable.

b. An orientation of a room or space to the outdoors to make the interior space feel larger. Aligning openings or using similar or sympathetic materials can help blend an interior space with an exterior space such as a patio or enclosed courtyard.

c. A physical connection or adjacency that opens up one or more areas to another.

Structure

Although structure may seem like just the utilitarian bones that help hold up roofs and floors, when expertly revealed and arranged, the structure defines space and suggests connections in very subtle ways.

Many designers will use structural necessities, such as bearing posts and beams, to create a delicate layering of line and shadow within a small space. This overlay can help define specific areas without the need for full-height walls, while still producing an overall feeling of openness.

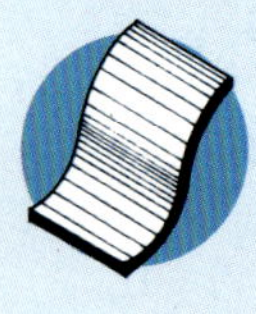

Surfaces

Modulation of wall, ceiling, and floor surfaces can produce rich and varied textures that enliven even the smallest room. Surface variations can also be employed to affect how light fills the room. For instance, a mottled or highly textured surface diffuses light; while a polished or lacquered plane reflects it. Surface treatments range from basic wallpaper and paint to such esoteric materials as stainless steel, custom paint finishes, and special woods.

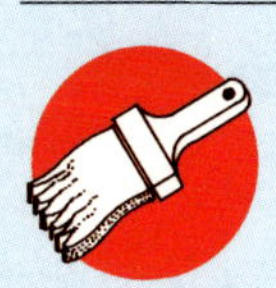

Color

Like surfaces, use of color can often make or break a design. In a small space, color can be used to highlight or de-emphasize a particular surface or object. Well chosen colors can be used to tie several small rooms together, or alternately, to subtly differentiate them. Color can be used to make surfaces recede or to draw them closer to the eye. Natural and artificial lighting greatly contributes to the perceived size of a room. Depending on the intention, a murky chiaroscuro or a bright, almost stark appearance can be used to open up a space or create a focus within a room.

Lighting

Natural light can be introduced unaltered via windows and skylights or it can be filtered and modified through deep recesses, window treatments or other means.

Lighting fixtures are currently available in an almost bewildering array of types and designs. Like natural light, artificial lighting can be used indirectly to merely suggest a warm glow or at the other extreme, the lighting fixture itself can be made a bright and dazzling centerpiece to the room.

Trim and Detail

Attention to details such as trim, connections, and hardware is important in all designs, but in a small space the importance of detail cannot be overlooked. Well-crafted details can suggest a richness that belies a small project budget. At the same time, details can introduce a perceptual scale change that camouflages the true dimensions of a small space.

Inside and Outside

Connecting inside and outside spaces is a time-honored device that is particularly appropriate to small space design. By sharing or borrowing space from the outside, almost any room can be made to feel larger and more gracious. The connection may be as simple as a lushly planted garden casually placed outside a pair of glass doors or as complicated as the rigorous use of similar or sympathetic materials that continue from indoors to the outside.

Furniture

Designers often employ built-in and carefully selected pieces of furniture to maintain and reinforce their aesthetic intent. In a small space, this aesthetic control is still important but furniture design and placement can, quite simply, save space—or at least use the space more efficiently. Striking material palettes in furniture can also be used to complement the background surfaces.

Not all of these design devices are appropriate for every project; using every device in a small project would be like ordering every item from an a la carte menu; the meal would simply be too large, too rich, and lacking the focus of a single well executed entree. In much the same way, most designers go through an editing process to try to achieve a simplicity and clarity of design. As you will see, each project illustrated in this book utilizes one or more of these design devices to make a small space useable and, at the same time, a delight to use.

Bedrooms

Too many times, "ideal" bedrooms are presented as huge, multi-use spaces that must, by definition, match the size of many people's entire living space. Nothing could be further from the truth; many times the best bedrooms are compact in footprint, efficient in organization and layout, and still appealing, imaginative or even provocative in their aesthetic.

The bedroom can be many things to many people. For some, the space is simply a place for rest, unencumbered by any other functional or practical consideration. In this situation, a bed and one or two pieces of furniture will suffice. Often a bedroom need be little more than an alcove that can be isolated from the remainder of the living areas for a degree of privacy.

For others, the bedroom is a retreat, a spot in which to sleep as well as relax and renew. For a child, the bedroom may be a retreat and also a play area, a space where imaginary adventures can begin far away from a parent's watchful eyes. In these cases, the sleeping functions are often subordinate to the need for relaxation or play—requiring a more complex and multi-layered design.

Sometimes, the demands of daily life will still impose themselves on the bedroom and the space must incorporate a small office area for bill-paying or letter writing.

In every circumstance however, any or all of these functions can be artfully incorporated into small spaces with little or no sacrifice in either convenience or delight.

SMALL BEDROOMS

Master Bedroom

This architect-designed bedroom uses simple devices of architectural elements and saturated colors to create a sense of expansiveness that extends throughout the house. Careful attention to movement through the spaces and to culminating vistas makes for an enriching and appealing visual and spatial experience.

Simple but thoughtful use of medium chroma colors on adjacent walls helps to break up the space into several planes, confusing the eye to make the room appear larger.

Photographer: Doug Hill

Photographer: Henry Bowles

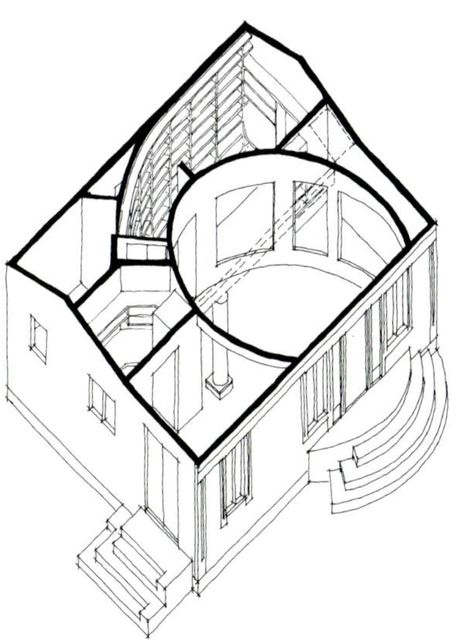

Alcove Bedroom

One bedroom shares the green color of the adjacent living space to suggest continuity between the small spaces.

Tightly organized, built-in shelving and a built-in bed maximizes the room's space while providing for all the necessary functional needs.

Photographer: Woody Cady

Attic Bedroom

A green that recalls the outdoors connects this low-ceilinged bedroom with the views beyond.

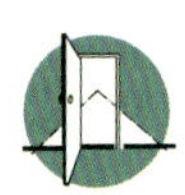

Shortened walls at the perimeter of the room, coupled with a shallow-set window and skylight, establish a deceptive visual datum line that lends the illusion of height.

Photographer: Cindy Linkin

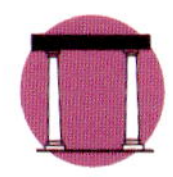

Exposed roof structure above lends a porch-like quality to the small scale reading area.

Use of the traditional blue porch ceiling color reinforces the outdoor quality of the space.

Classically inspired wood trim continues the spirit of the original house yet allows for differentiation in the small tower addition.

Bedroom Addition

A small reading area is located in the tower-like addition to a second floor master bedroom. While the main space accommodates bed and furniture, the reading area explodes upward with a tall ceiling that provides a spatial counterpoint to the lower-ceilinged bedroom area.

Photographer: Woody Cady

Photographer: Cindy Linkins

Bedroom Expansion

Restrained wood molding and trim profiles along with built-in shelving units recall the grandeur of large libraries.

A small combination library and sitting space off the master bedroom began life as a screened porch. Separated from the sleeping area by low walls and columns, the space provides an attractive focal point to the room beyond.

Photographer: Norman Smith

Studio Bedrooms

The pale cathedral ceiling lifts the eye upward in a small, plain bedroom, suggesting sky, open air, and light.

Furniture in soft colors blends nicely with the pastel ceiling, lending the space a sense of continuity without overpowering the room.

A bedroom overlooking a studio area is screened with sliding, translucent panels that borrow light from windows beyond.

Photographer: Paul Warchol

Photographer: Robert Perron

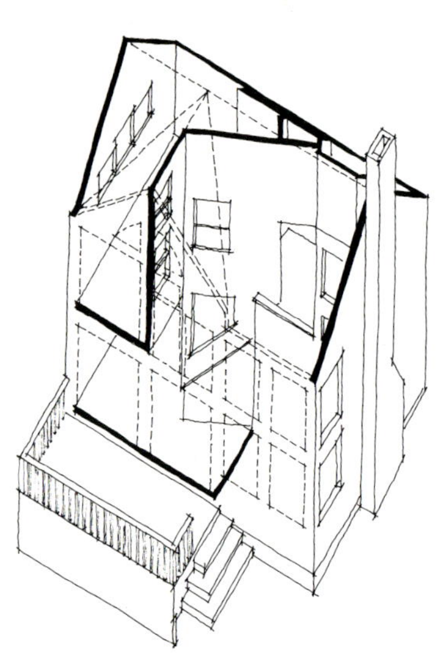

Loft Bedroom

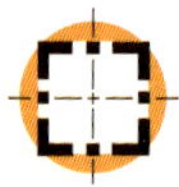

This stairway transforms into a small sleeping loft overlooking areas below.

Skylights above and windows on the inner wall make the stairway both a pathway and an important source of light for both this bedroom and small house.

Urban Apartment Bedroom

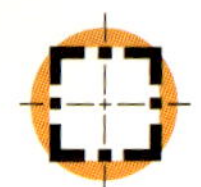

An unusual shared living and bedroom area maximizes a small space but allows separation via a sleek, curved wood wall. Closet and bathroom are also contained within a single, shared space.

Photographer: Andrew Bordwin

Open Plan Bedroom

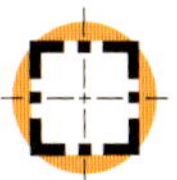

A bedroom open to living areas and fireplace can be closed off with blinds above the bed.

Wall openings of varying size allow indirect natural light to filter into the room.

Photographer: Robert Perron

Baths

The bath started life as simply a place in which to perform daily cleaning rituals. However, over the years and with changes in people's lives, those rituals and the bath itself have acquired a newly important and central place in daily life.

Like the kitchen, the bath is a space that must not only be functional but must also provide a restful and appealing environment. At the same time, the tradition of shutting the bath away from the rest of the house can be tempered in small spaces by design that honors the sybaritic nature of cleansing.

Small baths can provide pleasure, comfort, and practicality just as well as a large room, however they often have to work a bit harder to be successful.

The best small baths are well organized but maintain a measure of delight that transforms the mundane into the extraordinary. Since the small bath is normally *quite* small, the designer must also carefully edit the bath design and the devices that are used to shape the final product. Depending on the individual situation, one or more devices such as color, surfaces or lighting will help create the appropriate environment without overpowering the senses.

SMALL BATHS

Alcove Bath

Even the smallest area of a bath can be elaborated upon to good effect through the thoughtful juxtaposition of materials. Alcoves and other spatial manipulations can produce an impression of spaciousness without requiring vast amounts of floor area.

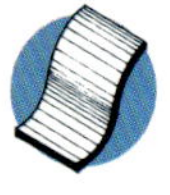

A mixture of masonry materials and wood lends a primal simplicity to this lavatory alcove.

A single pendant light provides a delicate focus to the materials and a pinpoint light source that balances the natural light.

Photographer: Undine Prohl

Remodeled Bath

As part of a more extensive bath remodeling, this alcove was created to clearly define the bathing area and to provide visual and spatial relief to the bath as a whole.

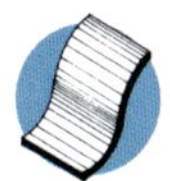

Mixing tile sizes and using pale colors to define horizontal planes helps to visually enlarge this bath tucked in an alcove.

An over-sized, round window brings a sense of whimsy as well as a more open atmosphere and a great view to the trees beyond.

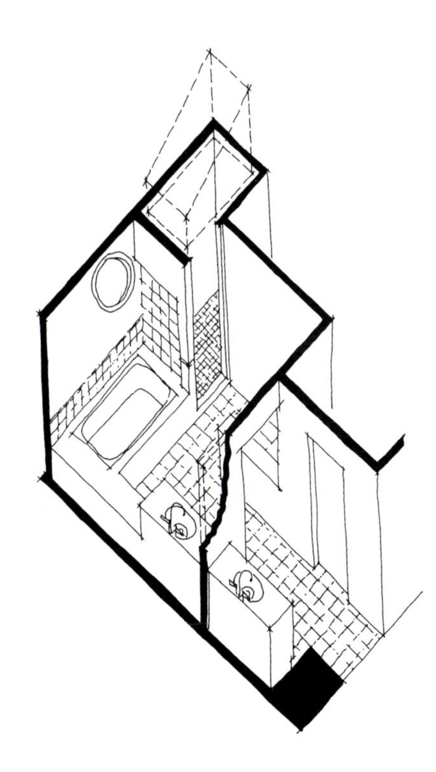

Photographer: Bennett Frank McCarthy Architects, Inc.

Japanese Apartment Bath

Natural wood contrasted with white walls creates an elegant yet meditative bathing area.

Shoji-like screens allow privacy with a diffused light.

Photographer: Richard Mandelkorn

Photographer: Undine Prohl

Open Plan Bath

Exposed wood posts and ceiling joists create a rich play of textures; they also help suggest smaller spaces within the small open room.

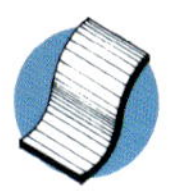

Natural wood ceilings and floors provide a surface continuity between adjacent spaces to expand their apparent size.

Walls that don't reach the ceiling divide the space without blocking light.

Half-Baths

An extremely saturated, darkly hued red enlivens this small sink area in ways that a lighter color could not.

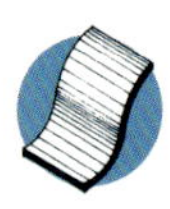

A simple use of floor to ceiling tile creates a visual and tactile presence in this small bath.

Necessary storage is tucked neatly into the corner.

Photographer: Woody Cady

Photographer: Richard Mandelkorn

Photographer: Bruce Katz

Classical Bath

Light value pastels with white trim and ceiling expand the space and help reflect natural light in this 80-square-foot bath.

Abundant natural light is introduced through windows and a door as well as via a skylight near the tub enclosure.

Classically inspired molding reinforces the meditative feeling in the room.

Gabled Bath

Color variations among bright tangerine walls, crisp, contrasting black tiles, and angled ceiling planes confuse the eye to make this bath appear larger.

A tub skylight illuminates the tub itself and helps to backlight the rest of the room, adding depth.

Mixed-Storage Bath

This small alcove is made into an attractive focal point by juxtaposing glass shelves above with closed storage below.

Photographer: Richard Mandelkorn

Photographer: Woody Cady

Photographer: Paul Warchol

STUDIO BATH

In this studio project, the bath area is not sequestered away behind solid walls but instead borrows light and a sense of views from the adjacent spaces.

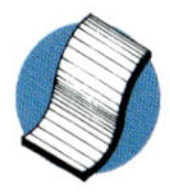

The sleekness of industrial materials is an appropriate choice for this studio bathroom.

A thoughtful reconsideration of the prosaic bathroom sink results in a provocative assembly that grabs the eye.

Photographer: Norman Smith

Master Bath

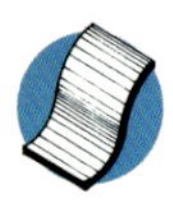

A random pattern of teal, yellow, and grey tiles on the floor and countertop disguises the small size of this master bath.

A custom designed, birch plywood vanity suggests individual spaces for two people.

Photographer: Norman Smith

HALL BATH

Simple grey walls in tandem with quartz-halogen wall lighting produce a variety of light and shade in a small powder room.

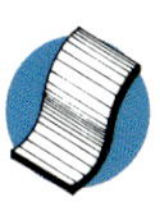

Glossy black floor tile makes the tiny actual floor area disappear.

Hallway Vanity

A master bedroom remodeling occupies the second floor of this Cape Cod style house. Rather than take precious space away from closet, storage, and the bedroom proper, the hallway does double duty.

A vanity area is tucked within the hallway connecting the master bath to the rest of the house.

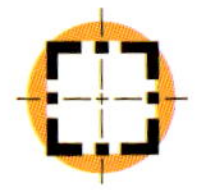

To create the minimum size for a vanity, this bath borrows space from a closet in the adjacent room.

Photographer: Norman Smith

Accessible Bath

This existing bath was updated and expanded to add light, as well as to accommodate a disabled user. A simple yet seemingly spacious layout allows access without sacrificing a sense of delight.

A tall sand-blasted glass window adds interest to a monastic design and bathes the room with diffuse light while preserving privacy.

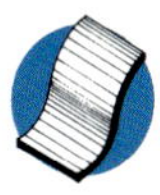

The shine of glass, tile, and a lacquered wood ceiling gives surfaces in this bath a sleek but low-key appearance.

Photographer: Norman Smith

Corner Space Baths

Tile color variations emphasize the corner window and draw the eye away from the irregularly shaped space. Glass block in a corner window spreads a soft, diffused light without sacrificing privacy.

Photographer: Woody Cady

Drapes hide a toilet niche in a corner bathroom wall.

Photographer: Norman Smith

Children's Bath

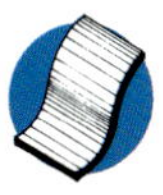

A clear-sealed, plywood floor with segments of stained inlay graphically divides the space to suggest different areas within a single small space.

Stained and painted wood trim extends the floor pattern up onto the walls and ceiling.

DIRECTORY OF ARCHITECTS

The Alexander Group, Inc. (builder) . . . . 32 left

Bennett Frank McCarthy Architects, Inc. 29

Bruce Bierman 22

Centerbrook Architects and Planners 20, 21

Frederick Fisher, Architect 28, 31

Heritage Building and Renovation, Inc. . . . . . . 34 right

Interim Office of Architecture 19, 35

James Mary O'Connor 14

Jay Davies/Architects at Work 40

Lindsay Boutros-Ghali, Architect . . . . 30, 32 right, 34 left

Moore Ruble Yudell Architects and Planners 15

Norman Smith Architecture 19, 36, 37, 38, 39

Robert Ford Architecture 23

Shorieh Talaat Design Associates 17, 18

Woodie Hartzell 16

DIRECTORY OF CONTRACTORS

Bensley Construction Company/Shozhan Woodcrafters 30

Lloyd Harrison & Associates 38

Lofgren Construction Company 33

Shorieh Talaat Design Associates 17, 18

Taurus Renovation & Construction/Inverness Builders 19, 36, 37, 39

William Kellar, Inc. 32, 34

DIRECTORY OF PHOTOGRAPHERS

Andrew Bordwin
70 A Greenwich Avenue #332
New York, NY 10011

Henry Bowles
933 Pico Boulevard
Santa Monica, CA 90905

Woody Cady
Woody Cady Photography
4512-A Avondale Street
Bethesda, MD 20814

Bennett Frank McCarthy Architects, Inc.
7003 Carroll Avenue
Takoma Park, MD 20912-4429

Douglas Hill
2324 Moreno Drive
Los Angeles, CA 90039

Bruce Katz
2700 Connecticut Avenue NW
Washington, DC 20008

Cindy Linkins
Shorieh Talaat Design Associates, Inc.
15715 Kruhm Road
Burtonsville, MD 20866

Richard Mandelkorn
65 Beaver Pond Road
Lincoln, MA 01773

Robert Perron
119 Chestnut Street
Branford, CT 06405

Undine Pröhl
1930 Ocean Avenue #302
Santa Monica, CA 90405

Walter Smalling
1541 Eighth Street NW
Washington, DC 20001

Norman Smith
3800 Military Road NW
Washington, DC 20015

Paul Warchol
133 Mulberry Street
New York, NY 10013